Copyright © 2012 by Anders Fjendbo Jørgensen

All rights reserved.

No part of this publication may be reproduced, distributed, or transmitted in any form or by any means, including photocopying, recording, or other electronic or mechanical methods, without the prior written permission of the publisher, except in the case of brief quotations embodied in critical reviews and certain other noncommercial uses permitted by copyright law. For permission requests, write to the publisher, addressed "Attention: Permissions Coordinator," at the address below.

Solar Health Aid Press
Brændekilde Væde Vej 36
5250 Odense SV, Denmark
www.healthaidplus.com

Printed in Denmark

Sterilization of Instruments in Solar Ovens.

By
Anders Fjendbo Jørgensen
Kirsten Nøhr
Flemming Boisen

INTRODUCTION

The sterilization of instruments has become an increasing problem especially after the appearance of AIDS. A common method to disinfect instruments was until recently use of chemicals, but this method can no longer be recommended. For heating, fossil fuels are expensive and not readily available in many places, and wood for fuel is becoming more and scarcer because of deforestation. Fortunately, the insolation is as high as 1000 W sq.m-2 in many developing countries. Therefore, it is relevant to utilize solar energy also for the sterilization procedures. One possibility is to use photovoltaic systems for generating power for autoclaves. This is a well-known technology, but it is still a very expensive and inefficient way of generating the necessary heat. The photovoltaic panels can only use 12-15% of the isolated energy, and the capital cost is too high in most places in the less developed countries. Furthermore, the presently known autoclaves use a rather sophisticated technology that cannot be manufactured in many developing countries, which leads to dependency on the industrialized countries.

The basic idea of the present research project was microbiologically to test a designed sterilizer that does not need fuel of any kind and that can be produced, installed and maintained by the local community where it is used. The sterilizer must be easy to use and reliable, must require only simple maintenance, and must be relatively cheap to produce from locally available material. A Maria Telke solar cooker was modified to suit this purpose.

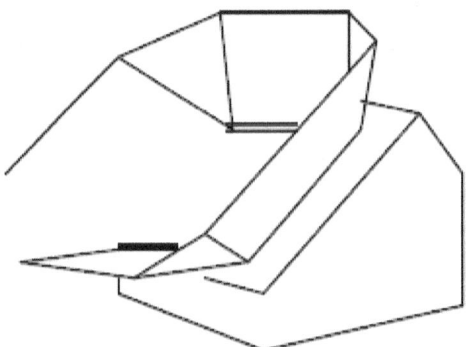

Figure 1. Telke's solar oven. A solar sterilizer for sterilizing instruments at rural clinics without access to grid electricity was developed from the original Telke solar oven. The sterilizer was tested microbiologically. A meter that monitors the sterilizing process has also been developed.

Hot air sterilization

Sterilization is the freeing of an article from all living organisms, including viable spores. Sterility is a theoretical concept that can never be reached with absolute certainty. In practice, however, a satisfactory sterilization process is one designed to ensure a high probability of achieving sterility, e.g. a process that provides more than 6 Log reductions of organisms (e.g. spores of bacteria) of a defined, exceptionally high degree of resistance. The commonly used test bacteria is *Bacillus subtilis var niger* with a D-value of 7.5 min and a Z-value of 25 (White 1987).

Findings from Tanzania have shown that bacillus spores isolated from a river in this hot climate are more heat resistant than isolates from temperate climates. An isolate of *Bacillus subtilis* (TN12) has a D-value of 16.2 min. (Jørgensen et. al 1998). This must caution one also to make tests with bacteria spores that are relevant for the place of operation. Thus, for the sterilization process to be satisfactory for hot climates, at least 97 min at 160oC (6 times the decimation time) is required. The time needed at other temperatures can be estimated from the Z-value of the test bacteria.

Normally for the test spores an increase of 25o C can reduce the time to one tenth (a Z-value of 25). Unfortunately it was not possible to determine the Z-value for the Tanzanian isolates because it might be that the Z-values are also higher for isolates from hot climates.

To be on the safe side we decided to assume a Z-value of 32.8 degrees C (this value implies that the time will be reduced by half every time the temperature is increased by 10oC).

This relationship between temperature and time allows us to calculate the sterilization effect when the temperature fluctuates as it does in a solar oven. The risk of surviving microorganisms after sterilization also depends on their number at the beginning of the process. It remains good practice to clean the instruments carefully before they are put in a sterilizer.

The effectivity of the sterilization can be tested by physical, chemical, and biological methods.

The sterilization process can be tested physically by a temperature record chart produced for each run; this is relatively easy to interpret when the temperature is regulated by a thermostat, but when the temperature fluctuates special calculations have to be employed. We chose the term "sterilization effect" to be the log reduction of numbers of viable bacteria spores.

The value 1 was given for 10 min at 1600 C, 2 for 20 min at 1600C, etc., and 2 for 10 min at 170oC, 4 for 10 min at 180o C, and 8 for 10 min at 190o. This gives the equation $f(x) = 0.000015259 \times 1.071773463t$ (1); $f(x)$ is a function of a period of 10 min sterilization and t is the temperature at the beginning of the period. By using a simple microprocessor, the temperature-time relation can be processed, and an integrated value for the sterilization effect can be given simultaneously; this is preferred here to a manual check by a chart recorder. Physical methods require the temperature to be measured at relevant points. Ideally the temperature sensor should be placed where the temperature is lowest, and on the surface of the instruments that are most difficult to sterilize.

The heat in a solar oven can be expected to be unevenly distributed, and lowest inside boxes. Chemical methods to indicate sterilization are available as Browne's tubes (Albert Browne Ltd., Leicester, UK). The tube fluid changes colour from red to green after a certain time period at a high temperature. According to the product data sheet, Browne's tubes almost obey the time/temperature relations as those described above and in the equation.

Browne's tube type 5 changes colour at approximately $f(x)=12$, and Browne's type 3 changes at $f(x)=6$. Browne's tube 3 is usually recommended when a temperature chart recorder is not available (White 1987).

The biological methods remain the reference standard. Ideally samples of the instruments should be tested for bacteria spores after each sterilization process, but this procedure will never be practicable.

An easier way is to have samples of spores relevant for the geographical location prepared in a suitable medium in, for example, sealed glass ampoules or sealed aluminium foil. The spores have to be in a carrier medium that does not reduce their ability for regrowth. The medium can be fine sand, clean and dry, without any chemicals or salts which prevent growth.

Special techniques have to be applied for the test organisms to sporulate, and for harvesting the spores (Jørgensen et al. 1998). In addition, the spores must be put at relevant places in the sterilizer.

MATERIAL AND METHODS

The solar oven

The chosen design was a Maria Telke solar oven, which was originally used to cook in (Telke 1955). This oven is reported to be able to generate temperatures above 180o C, which is sufficient for hot air sterilization.

The prototype was constructed of two aluminium boxes with 5 cm insulation material in between. On top is an opening covered by two layers of tempered glass and reflectors made of aluminium. The upper surface is angled at 30o to the horizontal so the maximal insolation will take place in the afternoon if the oven faces west. In most places this will be the best time for the sterilization because the peak work load is mostly in the mornings, and that is when the instruments are used. The items for sterilization can then be cleaned and put in the oven before it gets too hot, and the items can be taken out early the following morning when the oven has cooled down. To make the sterilizer more user-friendly it can be built into a wall of the clinic facing west.

The oven can then be opened from inside the clinic building, and the instruments can be put in without leaving the building. The solar oven was manufactured in Tanzania using locally available material.

The reflectors were also made of aluminium sheets to make the oven less vulnerable to theft and strain, even though the reflection is poorer compared to glass mirrors. An aluminium box (10 x 40 x 60cm) was made and painted black using an ordinary non-glossy blackboard paint. The instruments were placed in other small aluminium boxes inside the black box. The small aluminium boxes contained instruments for separate clinical procedures (such as repair of wounds, delivery, etc.) and dressing material, glass syringes, and needles made of steel.

Test of effectivity

The temperatures in the oven were mapped using a wire sensor and an electronic monitor (OI-electric, Denmark). The sensors were placed in different positions in the oven, in the large black box and inside the small aluminium boxes to detect differences in temperature. The prevalent temperature was recorded. To check the value of the insulation, the temperature on the outer surface of the oven was measured on a few occasions.

The temperatures were recorded every ten minutes during the procedures and displayed graphically using a computer. The testing procedures took place in October, which is a time of year with relatively high insolation and few clouds.

Microbiological testing

Two types of commercially available test organisms were used. *Bacillus subtilis var niger* (no 9372 ATCC) and *Bacteriun subtilis* (from the State Serum Institute, Copenhagen, Denmark). The organisms were in a ready-to-use glass ampoule or a metal foil, respectively. Spores from bacteria isolated from the Mlalakuwa river in Tanzania were also used. Each ampoule contained about 107 spores. The ampoules were placed in selected positions in the black box and in the small aluminium boxes, and also inside the glass syringes and inside the dressing material to simulate "worst case" positions.

The tests were performed using the available sun as the only energy source. When it became certain during the test period that sterility could be obtained, the time was reduced when the insolation was high.

A set of values for time and temperatures was obtained, and they could be compared with the results of the growth of bacterial spores. The methods for culture of the spores are described elsewhere (Jørgensen 1998). In order to find out whether any position was poorer than any other for sterilization, the tests were done in a systematic way by defining control places in the large black box, i.e. each corner and the centre. As a chemical check the different types of Browne's tubes were put in the same positions as the bacterial spores.

An electronic device to indicate sterility

The above mentioned mathematical formula was programmed into a microprocessor after the first tests. The microprocessor was then able to compute the signal from the sensor (a thermocouple), and a light diode switched on when the temperature/time factor was sufficient. At first the mathematical formula was programmed using an ordinary computer that was connected to the sensor via an interface. The outcome was then tested for a period of some months. The final model, called Bacto-II, was constructed later.

The electronic components were built into a small box, and the sensor (thermocouple), which can sense temperatures from 00 - 350 0C with an accuracy of about 1/2%, was put in a position with temperatures similar to the temperature of the material to be sterilized. Bacto-II was fitted near the oven door and connected via a separate hole.

Testing the device in a rural health clinic

The chosen clinic was Zinga Rural Health Clinic some 60 kilometres north of Dar es Salaam in Tanzania. The clinic has one Clinical Officer, one nurse, and two nursing assistants. The building consists of six small rooms. The sterilizer oven was placed outside the treatment room, and an opening in the wall was made from the room to the sterilizer. The oven door was fitted into the opening so that the oven can be loaded from inside the clinic. The inner oven door opened downwards to serve as a shelf for the box. To secure insulation in between the oven and the room, a specially made "pillow" was made from heat-resistant material.

During the test period the temperatures were recorded every ten minutes in a data logger (Tiny Talk, London, UK) with its own sensor. The data were then loaded into a computer at regular intervals of 10 to 14 days. Following installation of the oven and the Bacto-II, the method of sterilization became part of the daily routine of the clinic, and it was assessed by the staff and the researchers.

RESULTS

Results of development of the solar oven

It was possible to make the oven of local available material at a cost of about $100, and it was manufactured using simple tools. The weight of the oven is less than 50 kg and it is easily transported to the place of installation. The Maria Telke oven was designed to be moved for sun tracking, but for clinic use the oven has to be in a fixed position. Thus, the oven was made wider (in the north-south direction) and less deep (in the east-west direction) to secure sufficient heating capacity throughout the year.

The construction would have to be modified for places at different latitudes. The inner glass cover broke after some time and it was necessary to replace the glass with tempered glass. The available tempered glass reduced the transparency by some 10%

The physical tests

The air temperature at Dar es Salaam was relatively stable during the test period (270 to 310 C). The wind velocity was relatively low (about 3-6m/s). The results of the temperature mapping inside the oven showed different temperatures at different places, but after some time the temperatures became similar. The air temperature inside the oven was approximately the same as the air temperature inside the black box. The temperatures inside the small instrument boxes, i.e. the most relevant ones, were low at the beginning compared with the air temperatures in the black box, but after some time of heating up, the difference was only 1-20.

During the cooling phase the temperature in the small boxes became higher than in the larger box and when measured in sterilization effectiveness (according to a previously mentioned formula) the difference between the two was small. The overall maximum temperature reached was 189.70 C inside the black box and 1850 C inside the instrument case. The maximum temperature was reached after some 1 hour on sunny days, but of course the increase in temperature was slower on days with clouds (results not shown). On one fully overcast day the temperatures never exceeded 900C. Except for the day with full overcast it was possible to reach temperatures over 1600C.

Figure 2

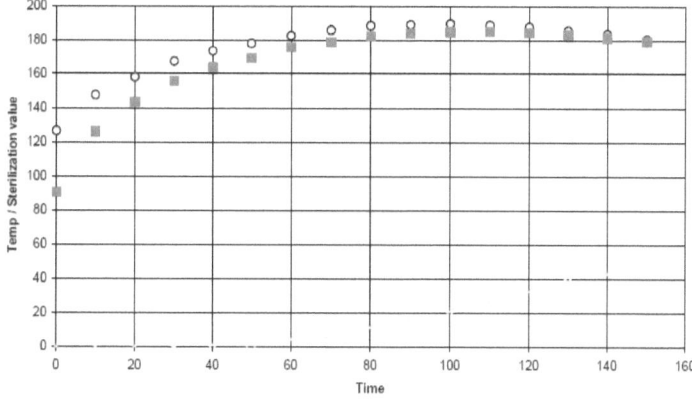

Microbiological and chemical tests

Any difference in sterilization effectiveness between the different places in the black box could not be detected by the biological indicators. With respect to the Browne's indicators, in some of the test type 5 turned green in the corners but not in the centre, indicating the sterilizing effect in the center might be lower than in the corners. Except for one test run, all the samples of spores were inactivated after the process.

One of the test runs was simulation of a "worst case" situation where the samples were put in towels to insulate the spore samples further. The test-run had a sterilization effect of 16, which was about the average for all the test runs. Also this test resulted in full inactivation of the bacteria spores.

Sterilization of instruments in a solar oven. Results of microbiological and physical tests

	Accumulated Sterilization effect value*	Maximum temperature degrees C	No. of Browne's indicator§ Type 3	Type 5	No. of spore samples with growth‖
1	0.02	$87°$	0/4	0/4	30/30
2	4.5	$168°$	2/4	0/4	0/30
3	6.4	$162°$	4/4	0/4	0/30
4	7.0	$177°$	4/4	0/3	0/30
5	10.5	$171°$	4/4	2/4	0/30
6	14.0	$185°$	4/4	4/4	0/30
7	15.9	$178°$	4/4	4/4	0/30
8	16.0	$170°$	-	-	0/41
9	27.0	$175°$	4/4	4/4	0/30
10	29.0	$178°$	4/4	4/4	0/30
11	51.0	$185°$	4/4	4/4	0/30

The results compared with the mathematical model

The results showed that the hypothetical mathematical formula corresponded well with the chemical indicators. Browne's type 3 changed color at a sterilization effect value of 6.4 or more, but at a value of 4.5 only two of the four changed color. Browne's type 5 was expected to change color for values above 12; it did change color for all values above 14, but for the value 10.5 only two of the four changed from red to green.

The tests with a sterilization effect value of 4.5 or more inactivated all the spores from the manufactured samples; this is better than expected because theoretically this value should only result in 3-4 Log reductions of the most heat-resistant spores isolated from the Mlalakuwa river. This indicated that the mathematical model was appropriate.

Testing the solar oven in a rural health clinic

The installation of the solar oven in the clinic was rather difficult. There was delay in the development of the temperature/timer meter (Bacto II), and the weather made it impossible to travel to the clinic in Zinga for long periods of time. During 1997/98 Tanzania experienced unusual rainfall and flooding.

The period was extraordinary due to the El-Niño rains and overcast weather for almost a year. This limited the solar oven sterilization to about once a week.
The oven was modified to allow more insolation during the extreme sun positions (December and June).
It was possible to fit the oven onto the clinic and for it to be used easily.

Bacto was supplied with electricity by solar PV-panels, which also powered lights in the clinic. For the Bacto II power supply to be secure when the battery charge was low, a battery control could switch off the light in the clinic to give preference to the meter. The oven was well protected against the rain, but water was leaking into the insulation material anyway and the oven had to be repaired.

The material for water protection oozed moisture, which condensed between the two layers of glass and reduced the transparency to a degree the oven was not working before the moist could be removed.

The concept of solar energy providing enough heat for an oven was new for the clinic staff. They got the impression that the electric power for Bacto-II was also the source of power for the oven. Thus, when the light changed for the first time from red to green they disconnected Bacto-II to save energy. The instruction material in English and Kiswahili had to be developed for the staff.

DISCUSSION

The development of the oven and its effectiveness

The results showed that it was possible to generate high temperatures in a solar oven, that this high temperature could be achieved in less than two hours, and that it was possible to keep the high temperature for a sufficiently long to secure sterilization.

We would have preferred a higher peak temperature at about 2000 C because the maximum temperature was too low in June-July and December-January. One possibility might be to equip the oven with three layers of glass but using the glass that is currently available in Tanzania would reduce the transparency further. Another possibility would be to increase the insulation of the oven. Further research is needed to solve this problem, without forgetting that the sterilizer has to be constructed from inexpensive locally available materials.

The microbiological tests showed that the solar oven could kill the vegetative bacteria even on overcast days, though all the bacteria spores were able to grow.

Direct sunshine is needed for the oven to sterilize. Then the prefabricated spores, both the usually recommended commercial ones and the more heatresistant ones originating from a hot climate, can be killed in the solar oven.

Growth is prevented after reaching a sterilization effect value of only 4 (equivalent to 1600 C for 40 min.), a value making only 2 of 4 Browne's type 3 change colour. The mathematical model proved to be in accordance with the Browne's tubes because the type 3 changed about the 4.5 value for sterilization effect and the type 5 between 10 and 14. The results of the microbiological tests, taken together with the D-values makes it difficult to choose the correct value for sterilization effect to secure complete sterilization. From the results for the prefabricated spores, the choice could be a value of 4.5 or higher. To be on the safe side, it was decided to have a cut-off value of 6.

After choosing the value for the sterilization effect, the electronic device was built with tree lights. The device was programmed to show a red light when the sterilization process is not done, the yellow to be on when the instruments are considered sterile for most purposes and a green light for values of perfect sterilization.

Use of the solar oven in a rural health clinic

Putting the solar oven into practical use proved a rather difficult step. It was not usual to think of the sun as something that can be used as a daily life resource. It was difficult to imagine that an oven could be heated by the sun to such high temperatures. The staff got the impression that the electrical power supply to Bacto II was also the source of energy for the oven but after receiving the instructions they very quickly adapted to the new way of sterilization, and the acceptability was soon high.

The installation of the oven was rather difficult because of the heavy rains, and because water was leaking into the insulation material. The results showed that the temperature was not sufficiently high outside the peak season.

The design of the oven has to be modified to increase effectiveness throughout the year. And a solar oven must not be the only method of sterilization in a rural health clinic.

CONCLUSIONS

It was possible to construct a thermal solar heated oven for sterilization of instruments that can reach temperatures above 180o C, and it was possible every day during the test period to reach temperaturesin the solar oven that were effective in killing all vegetative bacteria. During the test period it was possible on sunny days for the solar oven to fulfill the international recommendations for dry air sterilization. An appropriate device to indicate when the sterilization process has taken place was created.

The solar oven has proven to be a realistic method for sterilization of instruments in a rural health clinic. Further technical research is needed to optimize the solar oven for use during the whole year when the oven is in a fixed position.

References

Jensen, H.M. (1950). Håndbog i mejeriteknik. Mælk, smør og ost. ed. Jørgensen A and Jensen H.M. p 1087. 2nd edition, Alfred Jørgensens forlag, Copenhagen, Denmark.

Jørgensen, A.F., Nøhr, K., Sørensen, H. and Boisen, F. (1998). Decontamination of drinking water by direct heating in solar panels. *Journal of Applied*

Strandgaard, E., Jerpersgaard, P. and Grønbæk, O. (1987) In *databog for fysik og kemi* p 148. Copenhagen: F&K forlag.

Telke, M. (1955) Solar stoves. Transactions of the Conference on the use of Solar Energy, Vol. 3, Part 2. p 87-98. Tucsan, Arizona.

White, A.B.(1987) Sterilization in the laboratory, In *Practical Medical Microbiology*). pp 64-89.

www.ingramcontent.com/pod-product-compliance
Lightning Source LLC
Chambersburg PA
CBHW021857170526
45157CB00006B/2489